to John,
Our traveling companion, antique-hunting partner, and all around dear friend. thanks for exploring the world with us.

BACKYARD ARCHAEOLOGY
The Willis Allen House
1854–1994
With Family Histories of the Occupants

By
Bernard A. Paul
James Ross
Tracey Sandefur

Bernard A. Paul
J. Ross

Edited by Bonita K. Rubach
Illustration and design by Jim Balsitis
Word processing by Jayette Bolinski

Cover Photo: The Willis Allen House.
(Photographer: Jim Balsitis)

American Kestrel Books

I saw Mrs. Walsh at
Dookie Chase's. She ax about you.

Published 1994
Printed in the United States of America

00 99 98 97 96 95 94 5 4 3 2 1

Edited by Bonita K. Rubach
Illustration and design by Jim Balsitis

American Kestrel Books
American Resources Group, Ltd.
127 North Washington Street
Carbondale, Illinois 62901

Paul, Bernard A., James Ross, Tracey Sandefur
Backyard Archaeology
The Willis Allen House
1854–1994
With Family Histories of the Occupants

Includes bibliographical references.

CONTENTS

Preface .... *iii*

Acknowledgements .... *iv*

Introduction *1*

The Families *2*
- *Willis & Elizabeth Allen (1854-1859)* *2*
- *Dr. Augustus N. and Paulina Lodge (1860-1893)* *4*
- *Hiram and Mollie Boles (1893-1929)* *5*
- *George and Eliza Tyner (1929-1945)* *5*
- *Robert and Miriam Moore (1945-1964)* *6*
- *James R. and Lorene Wilson (1964-1973)* *6*
- *Bernard and Rebecca Paul (1973-present)* *7*

Archaeological Investigation *9*
- *Introduction* *9*
- *Summer Kitchen* *11*
- *Cistern* *17*
- *Well* *18*
- *Outbuilding* *20*
- *Trash Pit* *21*
- *Old Porch/Family Room and Pantry* *21*
- *Postmolds* *21*
- *Summary and Conclusions* *21*

Bibliography *22*

LIST OF ILLUSTRATIONS

Maps

1. *Location of the Willis Allen house* *1*
2. *Plan map of site excavation* *9*
3. *1922 and 1937 fire insurance maps* *11*

Plates

1. *Dr. Augustus N. Lodge* *4*
2. *Paulina Allen Lodge* *4*
3. *The Bernard and Rebecca Paul family* *7*
4. *East view of summer kitchen foundation* *12*
5. *Glass bottles* *13*
6. *Ceramic vessels and tableware* *14*
7. *West half of excavated cistern* *16*
8. *Displaced piers from outbuilding* *20*

Figures

1. *Painting of the front of the Willis Allen house* *10*
2. *Painting of the rear of the Willis Allen house* *10*
3. *East profile of cistern* *16*

PREFACE

The Willis Allen house, located in Marion, Illinois, was constructed in 1854 by Willis Allen, well-known prosecuting attorney, state senator, and U.S. representative, and is listed on the National Register of Historic Places. The house is currently occupied by Bernard and Rebecca Paul.

In September of 1992 Gail White, a close friend and business associate, introduced me to Becky and Bernie Paul. The Pauls were planning an addition to their historic 1854 home and had contacted Gail, an architect, to design their addition. He suggested that there may be important archaeological remains in the area of the planned addition and brought Bernie to my office to discuss the matter. The results of that meeting, the archaeological investigation carried out by American Resources Group, Ltd., and Bernie's history of the families that occupied the house from 1854 to the present, are the subject of this book.

I would like to thank Bernard and Rebecca Paul for their cooperation and enthusiasm, and Bernie's researching and writing the family histories; Gail White for introducing me to Bernie; and the authors and field party from American Resources Group, Ltd., Jim Ross, Tracey Sandefur, James Burrow, Gordon Howe, Bonita Rubach, and Charles Foor.

Michael J. McNerney

INTRODUCTION

On February 11, 1982, the National Register of Historic Places entered onto its listings the structure we know as the Willis Allen House at 514 South Market Street in Marion, Illinois (*Map 1*). Although much is known about this property and the personalities of those who called it home, much more needs to be discovered. Early in 1992, the archaeological study that is contained in this book was commissioned by Mr. and Mrs. Bernard A. Paul, the present owners of this historic home.

The first white settlers pushed into this area and began to carve homes out of the wilderness in the early 1800s. Milo Erwin said William Benson and his wife, Bethany, settled on "Poor Prairie," the present site of Marion, in 1826. They built a log cabin and cultivated corn and wheat. After they had established a foothold, the Bensons built a more elaborate cabin in 1835. They are reported to have kept a hotel of sorts in this house. It was also used as the first Williamson County courthouse in 1839.

Archibald Thompson Benson filed a land patent in the general land office in Shawneetown, Illinois on March 3, 1843, for those lands in the 33-acre tract south and east of the square, including what is now the site of the Allen House. In 1850 he and his wife Julia deeded that land to Willis Allen. The same tract was also deeded to Allen in 1854 by William and Bethany Benson. Congressman Allen paid $200.00 for each transaction. Barbara Burr Hubbs says Mr. Allen moved into the log cabin on the former William Benson land.

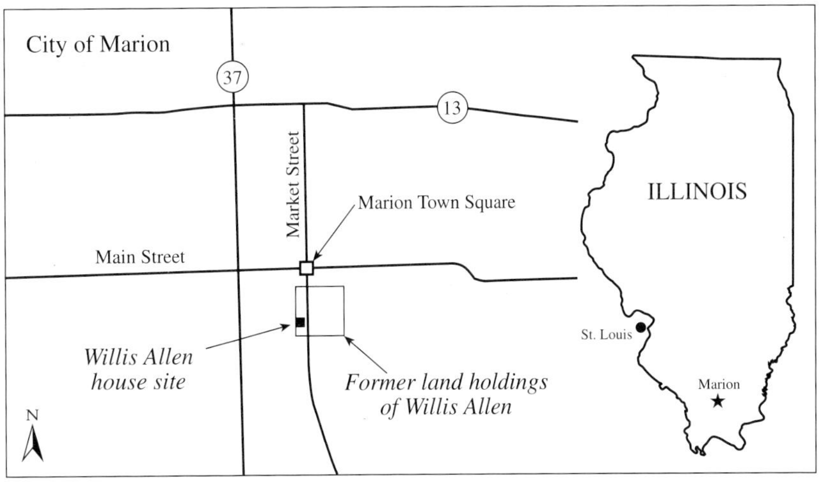

Map 1. Location of the Willis Allen house site in Marion, Illinois.

THE FAMILIES

Willis & Elizabeth Allen (1854-1859)

Willis Allen was born in 1806 in Wilson County, Tennessee. He married Elizabeth Joiner who was three years younger than he. Over the years of their marriage they had seven children: William Joshua "Josh" - born 1829; Jefferson Allen - date of birth unknown; Paulina Lodge - born 1834, wife of Dr. A. N. Lodge (the story of their occupancy of the house follows); Josiah J. - born 1838, died in childhood; Robert M. - born 1840; Marion Alderman - 1874-1876; Nannie Goddard Hendrickson - born 1843, according to the 1905 Souvenir "the first girl born in Marion," married first to J. Monroe Goddard, then after his death to Hartwell Hendrickson, the Circuit Clerk; Harriet Hundley - date of birth unknown, wife of Col. Robert M. Hundley (commander of the 128th Illinois during the Civil War; Mayor of Marion, 1866-69; and unsuccessful candidate for state senate, 1869).

Willis Allen moved to this area, then part of Franklin County, in 1830. He settled near Crab Orchard. He was politically active from his arrival and was a lifelong Democrat. He was elected Franklin County Sheriff in 1834. In 1838 he ran for the General Assembly on the platform of making a separate county out of the southern half of Franklin County. He was elected state representative and served in Vandalia along with Abraham Lincoln. That legislature, the last to sit in Vandalia the year before the state capitol was moved to Springfield, passed the Act dividing Franklin County and creating Williamson County.

When the new city of Marion was platted and sold, Allen paid twenty-three dollars for the lot west of the present Bank of Marion drive up facility on West Union Street. Dr. James Richey currently practices dentistry in the office located on that site.

Willis Allen and his family moved to Marion shortly thereafter. The 1840 census shows that his family consisted of six males and four females. Willis was then 33 and Elizabeth was 31 and pregnant with Robert. Josh was 11 and Jefferson was "five to ten," according to the census language. Paulina was six and Josiah was two. It is uncertain who the other two males and two females were who were then living with the Allens. Perhaps they had servants living in separate quarters around the farmstead, who were listed in the census along with the family.

Willis Allen served as Marion's first Postmaster in 1840. He was also elected Williamson County Prosecutor and held the office from 1841 to 1845. Attorney Allen was reelected to the Illinois legislature as a state senator in 1844 and again in 1846. He was selected to be a member of the State Constitutional convention in 1847. During this period he owned and operated several farms in the area which now comprises the southeast part of the city of Marion. He was named the first President of the

Williamson County Agricultural Society, (commonly known as the County Fair Board). Allen became a Master Mason and charter member of Fellowship Lodge #89, A. F. & A. M. in Marion, chartered in 1850. He served as its first Worshipful Master.

And most importantly Willis Allen was nominated for and elected to the U.S. House of Representatives in 1850. He was the first resident of Williamson County to be so honored by his fellow citizens. He served in the Thirty-Second Congress and was reelected again in 1852 to be in the Thirty-Third Congress.

It is likely that it was during this time of great political and financial success that he built his new home south of the square in Marion. Factors seem to indicate it was built about 1854.

The building is an ell-shaped, masonry-bearing wall, two-story detached dwelling with an entrance stair hall, constructed in the Italianate style.

The exterior walls are four withe brick masonry with interior walls of three withe masonry set on sandstone foundation. The bricks have been laid in common bond with three center brick rowlock arches over the windows on the second floor and soldier course arches over windows and doors on the first floor. The windows are long narrow double hung divided light sashes with sash boxes and sash wiights.

The roof style for the main portion of the dwelling is hipped with an extended cornice, embellished with cavetto moldings, frieze boards and ornate three-point tudor arch braces, installed in pairs. The roof style for the rear ell portion of the dwelling is a steep pitched gable roof with gabled dormers.

The staircase has walnut banisters and posts, fastened together with walnut pegs. Two functioning fireplaces have been retained downstairs.

Willis Allen served as teacher and mentor to young men wanting to "read the law." Such men as John A. Logan, Robert G. Ingersoll, and Allen's son, William Josh Allen, were his students. Classes consisted of the master presenting each student with cases, either real or hypothetical, and the student preparing his course of action, or "brief" in each situation. The students gathered to hear the senior Allen explain the "correct" answer in each case. When satisfied that the student had demonstrated sufficient scholarship and was competent to represent clients, the tutor vouched for him before the Supreme Court of the state. If found to be of good character, the Supreme Court admitted the new lawyer to practice.

In his 1876 history of Williamson County, the author and historian Milo Erwin paid homage to the renowned statesman:

He (Allen) was a man of considerable talent, great shrewdness, and unbounded energy. He lived respected by all, and idolized by his party . . . Allen was a spare built man, erect, graceful, and of uncommon strength, agility and endurance. His voice was soft and melodious, countenance ruddy and fair, his hair was russet brown, a soft grayish blue eye, lighted to a fascinating smile. He was frank, generous and confiding to a fault, and was more interested in doing a kindness to others, than serving himself. When he was engaged, his glance was withering. He . . . was the most powerful politician in southern Illinois in his day (Erwin 1876).

Willis Allen was appointed Circuit Judge on March 2, 1859. In this capacity he "rode the circuit," traveling from county seat to county seat, hearing cases wherever a docket had been scheduled. It was while holding court in Harrisburg that this highly successful and respected man died, on April 19, 1859. He was buried in his family cemetery behind the house on South Market Street in Marion.

But even then, history was not finished with Judge Allen. In 1904 the city of Marion was experiencing an epidemic of communicable disease. It was deemed a good public health measure for all cemeteries to be removed from inside the city limits. The Allen family cemetery was dug up and Judge Allen's remains and those of his other family members were re-interred in New Rose Hill Cemetery. His tombstone can be seen there, not far from the Goddard Chapel (presently the only other Marion structure to be listed on the National Register of Historic Places).

Plate 1. Dr. Augustus N. Lodge.

Plate 2. Paulina Allen Lodge.

Dr. Augustus N. and Paulina Lodge (1860-1893)

Augustus N. Lodge (*Plate 1*) was born in 1831 in Ohio. As a young man he enlisted in the U.S. Navy and served at Vera Cruz during the Mexican War. Following his military service he studied medicine, then moved to Williamson County.

Doctor Lodge married Willis Allen's daughter, Paulina (*Plate 2*). After Willis Allen's death the couple bought the Allen house from Paulina's father's estate. Dr. Lodge established a successful medical practice which he ran from his office in a separate building located on what is now the driveway of the Allen house.

The pawpaw trees which still grow around the back of the house and bear sweet fruit in the summer are credited to Augustus Lodge. It is said that he sat on the back porch eating pawpaws and threw the seeds into the backyard where they took root.

He was elected Williamson County Superintendent of schools in 1869 and served until 1877. In 1882 he was an unsuccessful candidate for the state legislature. He also served as Secretary of the Williamson County Agricultural and Mechanical Fair.

Mrs. Lodge died December 20, 1891 of paralysis. In 1893, Dr. Lodge sold the Allen house to Nelson and Mollie Boles.

He died in 1895. It is believed that he was buried next to his wife in the Allen family cemetery behind the home. His remains were later exhumed along with those of the other occupants and in 1904 were reburied at New Rose Hill Cemetery. His tombstone says, "He fed his fever patients." No one has an explanation for the significance of that cryptic memorial.

Dr. Lodge's daughter Stella married Edward Mitchell. Dr. Lodge delivered her twins, born in the upstairs back bedroom of the Allen house, on July 2, 1888. The boy, Dan L. Mitchell, was a lifelong resident of Marion and served in the U.S. Army in World War I. He never married. He passed away in the late 1980s. His twin sister preceded him in death.

Hiram and Mollie Boles (1893-1929)

Little has been recorded about Hiram Nelson Boles, nicknamed "Nellie," except that he was born in 1853 in Williamson County and was a saloon-keeper. He ran an establishment on the southeast corner of the Public Square in Marion, where the Hudspeth Building and the Stylart shop are located today. Nellie Boles was apparently a large man of large appetites, the image of the Edwardian era in which he lived. His picture appears in the 1905 Historical Souvenir of Williamson County as a member of the Agriculture Society. He died of "apoplexy" and alcoholism on August 3, 1912. He was 59.

By way of contrast, Hiram's widow, Mollie, was much better known. With her good looks and outgoing personality, she became something of an institution in Marion. Born February 8, 1869, in Rockwood, Randolph County, Illinois, Mary Elizabeth Tyner married Hiram Boles and moved to Marion early in "The Gay Nineties." They bought the Allen house from Doctor Lodge on May 2, 1893.

Her picture in the 1905 Souvenir shows Mollie to be a smiling woman of full figure. Several people have said that Mollie was the most beautiful woman in Marion. Many older residents of Marion still refer to the Allen house as "the Mollie Boles house." Mrs. Hosea Ferrell says that she knew Mollie and visited her often to do quilting in the upstairs bedroom of the house.

After Hiram's death, Mollie was in trouble financially and had to sell the house to pay debts. As Administratrix of her husband's estate, she sold the house at public auction on August 20, 1920, to banker George H. Goodall for $750.00. Fortunately, Mollie was able to buy it back from Goodall in October of that same year.

Unfortunately, scandal tarnished the Mollie Boles story. On the night of July 12, 1929, the whole town found out about Mollie's alleged affair with a married man, a prominent mine manager. It is said that she and Mr. J. M. Rollo had been parked out on a rural road in his car. While in the supposed throes of passion, she apparently suffered some sort of seizure. The hapless Rollo brought her into town for medical attention. But it was too late. Mollie was pronounced dead upon arrival, of unknown causes. Local gossips had a field day. A coroner's inquest was held, but the records were suppressed. They cannot be found today.

The Allen house was listed in Mollie's estate and valued at $15,000.00. The Marion First National Bank was appointed administrator of her estate, but made no distribution of her assets, probably because the bank failed during the Depression. Many years later George Tyner petitioned to be appointed the successor administrator of her estate. No final account of his stewardship of Mollie's assets is on file in her estate.

George and Eliza Tyner (1929-1945)

Mollie's brother was George Tyner. He was born July 13, 1874 in Carbondale. He was a coal miner. He met Eliza Lane Sylvester, the widow of a Carbondale pharmacist, and they were married sometime after Mollie Boles' death in 1929. Eliza Tyner, called "Tyzie," is remembered as wearing very thick lenses in her eyeglasses.

Mr. and Mrs. Tyner are believed to have installed the first indoor plumbing in the Allen house. A partition wall was constructed in the upstairs center bedroom to create a master bathroom. The remainder of the bedroom became a central hallway. An old newspaper was found under the floorboards of the upstairs bath when it was remodeled in 1975. The old, yellowed and very fragile pages bear a date of 1934.

George died on June 10, 1936, from myocarditis and influenza. He is also buried in Marion's Rose Hill Cemetery.

After George's death, Eliza Tyner made an apartment upstairs and rented it to regular tenants. Belle Cummins Lane was married to Will Lane, Eliza's brother. Belle's sister, Anna Cummins Woodside, came to live in Eliza's house in the upstairs apartment. Anna, a very small woman, had been the wife of Doctor W. W. Woodside of Carbondale. The good doctor divorced her in 1935 and married his secretary. Mrs. Woodside was left virtually penniless, dependant upon welfare. Her granddaughter, Cleta

Thane, remembers visiting her often and sitting out on the front porch of the home with her.

Mrs. Tyner's daughter, Libby, married Leborn Locklar who ran an appliance store located on the northeast corner of College and Market streets. The Locklars were recorded in the 1941 Marion City Directory as living in the Allen house along with Eliza Tyner. Jim Parker recalls visiting the Locklars with his father, and playing basketball behind the house with the Cox boys, Trevor and Gordon, who lived next door.

The house was allowed to deteriorate badly during those years following the Great Depression. Gene Hilliard said that in 1944 the Tyners offered to sell the house to him for $4000.00. He refused the offer. Eventually, Eliza, along with Josie and Robert Tyner of Washington, D.C. deeded the house to Robert and Miriam Moore on September 19, 1945.

Robert and Miriam Moore (1945-1964)

The house was in very bad shape in 1945. It was overgrown with vines and the front porch was dilapidated. Some windows were broken and birds nested inside. The old floors were no longer safe or serviceable and the walls were buried under several inches of old wallpaper. The fireplace in the living room had been covered over. The old outbuildings were falling down.

It was this mess that Robert and Miriam Moore took upon themselves. Mr. Moore ran a jewelry store on the ground floor of the Hotel State building on the square in Marion. "People thought we were crazy," Moore once told Bernard Paul. In spite of the obstacles, the couple moved into the Allen house and began the first phase of the renovation which continues to the present.

The one story wooden front porch which had extended the width of the east side of the house was torn off, the sheds and outbuildings were razed, and the vines were pulled down. The old handmade windows were removed, reglazed and repainted, and fitted with new springs and slides. A two-car garage was constructed at the end of the gravel driveway. The exterior of the entire house was painted white. It was destined to remain that color for the next 45 years.

Inside the floors were replaced with polished oak hardwood strips and the walls were replastered. The living room fireplace was repaired and a modern brick mantel was built. A bath tub and toilet in the kitchen were removed and the kitchen was modernized. The kitchen fireplace chimney became the flue for an oil-burning central heating unit. This new furnace and a hot water heater were installed behind a door and a new brick wall on the west end of the kitchen. The kitchen ceiling was lowered to hide heating ducts and plumbing pipes running to the other rooms. Ceilings upstairs were also lowered to improve the heating efficiency of the new furnace. A Victorian-era gas light chandelier in the front bedroom is still connected to the old city gas supply line running across the attic floor.

In 1964 the Moores decided to retire to Arizona. They sold the house to Mr. and Mrs. Jim Wilson.

James R. and Lorene Wilson (1964-1973)

Jim Wilson had been a popular funeral director in the community of Zeigler for many years. In 1953, he sold that establishment and relocated to Marion. He first bought the Maller Funeral Home at 907 North Market Street and renamed it the Wilson Funeral Home. Then in 1955 he purchased the Frick Funeral Home at 104 South Van Buren Street and merged the two services into the Wilson-Frick Funeral Home. Initially Mr. and Mrs. Wilson lived in family quarters upstairs over the funeral home.

Jim Wilson was born in West Frankfort, the son of the highly-respected principal of the Frankfort Community High School. He married the petite and lovely Lorene Bryan in 1933 and they were the parents of three children. Their oldest, James Bryan Wilson, born in 1936, became a Chief Petty Officer in the U.S. Navy, serving more than 20 years as a crypotologic technician aboard some of the Cold War's most advanced submarines and aircraft. Robert A. Wilson, born in 1939, became a funeral director and continues to operate the Wilson Funeral Home in Marion. Their youngest child, Susan Elizabeth, born in 1943, is currently in real estate sales in Albuquerque, New Mexico.

Mrs. Wilson put her signature on the Allen house with several significant changes. She converted the screened-in back porch into an enclosed family room. A powder room was added in the downstairs pantry area between the library and the family room. She installed large closets in the back bedroom on the second floor. Large numbers of paving bricks were laid behind the house to form a patio area for plantings and outdoor living. Mrs. Wilson filled the yard with beautiful garden plants. Her azaleas were a spectacular spring show. Inside, her collection of antique furniture was of museum quality and made a visit to the Allen house a memorable event.

Bernard and Rebecca Paul (1973-present)

During the Wilson's residency in the Allen house, Bernard Arthur Paul (*Plate 3*) was often a guest in their home. Paul had been a classmate of the Wilson's older son Bryan at Marion High School and the two were lifelong friends. While visiting with the Wilsons, Paul made the frequent comment, "If you ever want to sell this place, be sure to let me know." In 1973, Jim and Lorene took him at his word and offered to sell him the Allen house. Paul, struggling to become established as an attorney in private practice in Marion, made the financial leap from small subdivision house to large family home, and took title to the home in May of that year.

On December 21, 1974, the living room of the Allen house was the scene of a candlelit Christmas wedding between Paul and Rebecca Jane Baker. The Rev. Everett Lynch of Zion United Church of Christ pronounced the vows which Paul had written especially for the ceremony.

Rebecca Paul had frequently passed the Allen house in her childhood. She remembers remarking to her mother, "I wish I knew who lived there so that I could go visit them." Her wish came true. As the mistress of the house, she became a force for restoration that still dominates the Allen house today.

In 1974-1975, Mrs. Paul directed the redecoration of the interior of the home. Ceilings which had been lowered during the Moore's occupancy were restored to their original height. Closets ("telephone booths") which had been added in corners of the bedrooms were removed. A new closet wall was added to the north side of the front bedroom. The second floor area was reopened by the removal of the doors which had been added in the 1930s to make the upstairs apartment. The upstairs bathroom was torn out and extensively remodeled. While installing the new bathtub, the Pauls' contractor, Walter Pape, had the unique experience of accidentally thrusting his entire leg through the ceiling of the dining room below.

In 1989 the exterior of the Allen house was stripped of its many coats of white paint and the original terra-cotta colored brick was restored. A waterproof sealer was applied to help preserve the soft, handmade bricks. The many layers of worn-out roofing were removed and the old cedar shingles and sawn slab wood decking were replaced. Modern roofing materials were installed.

In 1991, R. Gail White, of White and Borgognoni Architects, in Carbondale, was hired to begin design of a family room/basement/garage structure to be added to the house. In the early spring of 1992, the Pauls hired William "Butch" Jack of Goreville to be general contractor for the new addition.

Plate 3. The Bernard and Rebecca Paul family.

The new addition was designed to be compatible with the architecture of the original Italianate house. In terms of proportions, materials, and architectural detailing, the 1992 wing blends with and matches the 1854 structure.

The planform of the Paul addition is approximately square, reflecting the shape of the original house. The bearing structures of the addition are higher than today's usual construction, in order to match the proportions of the older building. The new addition is offset from the house so that it doesn't distract from the overall appearance of the historic section. Its roof slopes are designed to blend with the original. Window styles and sizes, as well as the curved window heads, are duplicative of those in the main structure. The modern convenience of an attached enclosed garage has been added to the basement, but located facing the rear of the property to keep it invisible from the street. A Victorian garden house opposite the north entry area is used to balance the scale and to provide visual screening of the actual extent of the addition. The clapboard breakfast room which serves as a link between the old house and the new wing is stepped down in height to provide a visual transition between them. The all-glass conservatory room is an extension of the new sitting room reaching out to the exterior terrace. It helps to scale the height of the new addition to the natural contours of the property.

The Pauls are fortunate to be able to incorporate some of the history of another Williamson County landmark into the new addition. The Jordan Curve Store, established east of Crab Orchard nearly one hundred years ago, was about to be torn down to make way for the relocation of Route 13. Located not far from their headquarters at Shady Rest, the Jordan store had been a frequent stop for the notorious Birger gang, southern Illinois' legendary gangsters of the Roaring Twenties. With the kind permission of the Jordan family and Amax Coal Company, the Pauls' carpenters salvaged the pine flooring, the beaded board walls, some old doors and other materials from the historic store. These were cleaned and trimmed to size and installed in the new structure. People walking across the Pauls' family room can honestly say they are walking where Charlie Birger's feet once trod.

Similarly, the red tile floors in the lower level rooms and the garden house have a unique history of their own. Thanks to the generosity of chief brickmason, John Aldridge of Eldorado, Illinois, the house was able to obtain these dramatically beautiful and functional tiles. They came from the roof of the old Louisville and Nashville Railroad depot which formerly stood in the community of Maunie, in White County, Illinois. That building, probably constructed well before the turn of the twentieth century, was dismantled in 1955 by Mr. Aldridge, his father-in-law and brother-in-law. The tiles were kept stored by Mr. Aldridge until presented for use in the Willis Allen House in 1993. The Paul family is very grateful for his kindness, as well as for his superb craftsmanship.

The landscaping plan was researched and designed specifically for the Willis Allen House by Max Jones, of Plantscape Nursery in Herrin, Illinois. Mr. Jones and his talented crew have attempted to accurately find and place those species of plants that would be typical of plantings around an 1850s southern Illinois household.

ARCHAEOLOGICAL INVESTIGATION

Introduction

Written records and informal interviews provided an enormous amount of information concerning the occupation of the house. Tax records indicated improvements made to the structure, while deeds identified the chain of title of the property. Neighbors provided valuable insight about features of the home they remembered dating back to the 1930s. All of this information provided a context by which to interpret the findings of the archaeological excavations.

The initial excavations were exploratory in nature. The objectives were to locate additional information on the original Allen homestead, investigate changes in the use of the property and house by the Allens and subsequent owners over a period of one-hundred and forty years, and preserve this information before it was destroyed by construction activities. As is often the case in archaeological investigations facing time and financial constraints, a backhoe was used to remove the topsoil. This procedure may expose subsurface features for further intense excavations.

Approximately four inches of topsoil over an area of approximately 225 square feet was removed directly west of the fireplace, the suspected location of the summer kitchen. With the removal of the topsoil a total of 38 test units ranging in size from 2 ft. by 2 ft. to 2 ft. by 10 ft. was excavated across the site. This resulted in the identification of the summer kitchen, cold cellar, cistern, well, outbuilding, sandstone foundation stones, trash pit, pantry/restroom, and post holes (*Map 2*). Archaeologists refer to these structures and holes or pits as features.

Not all of these features existed during the Allen's occupation of the home. Evidence in the form of plat maps, paintings, and diagnostic artifacts provided information about which features were probably used with successive occupations.

Archival research involving the careful examination of turn of the century Sanborn Fire Insurance Maps failed to identify a summer kitchen or cistern. Consequently we know that by the 1900s, these fea-

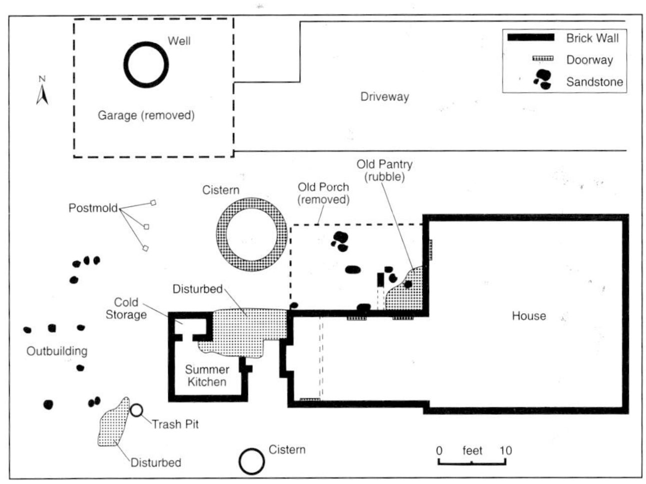

Map 2. Plan map of site excavation.

Figure 1. Painting of the front (east side) and the south side of the Willis Allen house, c. 1930s. By Rachel Williams.

Figure 2. Painting of the rear (west side) and north side of the Willis Allen house, c. 1930s. By Rachel Williams.

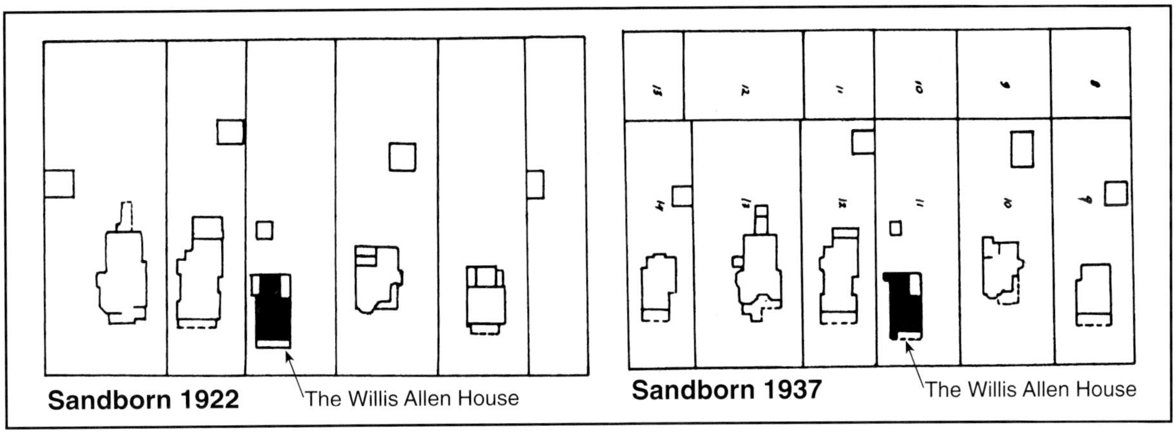

Map 3. 1922 and 1937 fire insurance maps of the property (Sandborn Insurance Company).

tures were no longer standing or in use by the occupants. This was reinforced by a painting made of the Allen House from a photograph dating from the late 1930s which did not depict the kitchen or the cistern (*Figures 1 and 2*).

Documentation of the manufacture of certain kinds of ceramics, glass, and architectural items has resulted in a series of manufacture dates historical archaeologists can use to date sites. Certain decorative techniques used on ceramic wares are diagnostic of a particular period of popularity. For instance, earthenware dishes decorated with a blue shell-edged pattern were popular in the first half of the nineteenth century. If an undisturbed feature is excavated that contains sherds from such a dish, then it is likely that the feature dates to that period. If, however, more recent artifacts are recovered from the same context (that is from the same fill within an undisturbed feature) then the feature must date at least to that more recent period. It is rarely the case that archaeologists find historical features which have only glass bottles and ceramic dishes with manufacturing and decorative techniques and patterns which are diagnostic of a short time period. A useful analytical tool used by archaeologists is the calculation of mean ceramic and glass dates. Each ceramic and glass artifact that has a diagnostic manufacturing or decorative technique has its own respective date. These dates are then combined and averaged by the number of glass and ceramic artifacts. The result is a rough estimation of the average date of the glass and ceramic artifacts respectively. This provides the archaeologist with a fairly good idea of when the feature was used.

Diagnostic artifacts were used to date the features identified during the archaeological investigation. Based on this information, it appears that two features date to the Willis Allen and Augustus Lodge occupations, one to the Augustus Lodge occupation, and one to the Boles occupation. Features like the sandstone foundation stones, pantry, post holes, and outbuilding are more difficult to date due to the relative absence of diagnostic artifacts. The outbuilding, however, is clearly marked on both the 1922 and 1937 Sanborn Fire Insurance Maps (*Map 3*). It is also interesting to note that the summer kitchen is clearly not on this map thereby suggesting that the structure was not standing at that time.

Summer Kitchen

The foundation of the summer kitchen was identified almost immediately following the onset of the backhoe excavation. The kitchen measures 12 ft. by 12 ft. and is situated approximately seven feet west of the southwest corner of the house (*Map 2*). The foundation is constructed of handmade brick identical to that used in the construction of the existing house (*Plate 4*). Much of the northeast corner of the summer kitchen is gone, due to twentieth century

Plate 4. East view of summer kitchen foundation and cold storage facility.

water line excavation. The foundation is double-brick wide and appears to turn toward the fireplace approximately four feet north of the southeast corner of the kitchen. This may have marked the point where the kitchen attached to the fireplace or, perhaps, the doorway of the kitchen where food was transported back and forth from the kitchen to the fireplace. A feature measuring six feet east/west by three feet north/south was documented in the northwest corner of the kitchen. The area's outer limits were bounded by a double-wide brick wall that extended a foot below the brick foundation. This feature was interpreted to be a cold storage facility probably accessible by a door in the floor of the kitchen (*Plate 4*).

The probability of finding a detached kitchen at this location was quite high due to the presence of a fireplace with a bricked up opening facing the outside (*Figure 2*). Detached or summer kitchens were common features of nineteenth century homes located in warmer climates. During the summer months all of the cooking activities were conducted in these structures in order to remove the heat from the main house.

Artifacts recovered from within the structure support the summer kitchen interpretation. Extensive amounts of butchered pig and cow bone were recovered throughout the structure. Pork and corn were the more popular choice of food for the settlers of southern Illinois who came from the southern region of the country. This cultural tradition is referred to by archaeologists as the Upland South Tradition. The life ways of this tradition originated in the eighteenth and nineteenth centuries in the Middle Atlantic region of the United States and later spread through Kentucky and Tennessee to southern Illinois with the western migration. The Upland South Tradition is generally characterized by a diversified farming and wood technology. Every family member at an early age plays an important role in running and maintaining the farm. County seats run by a small faction of persons and a stratified social system with slaves are also characteristics of this tradition. Of course slavery was not present in Illinois.

In addition to the pig and cow bones recovered, cinders suggestive of cooking activities were also documented throughout the structure fill, although they appeared to be most concentrated in the south-

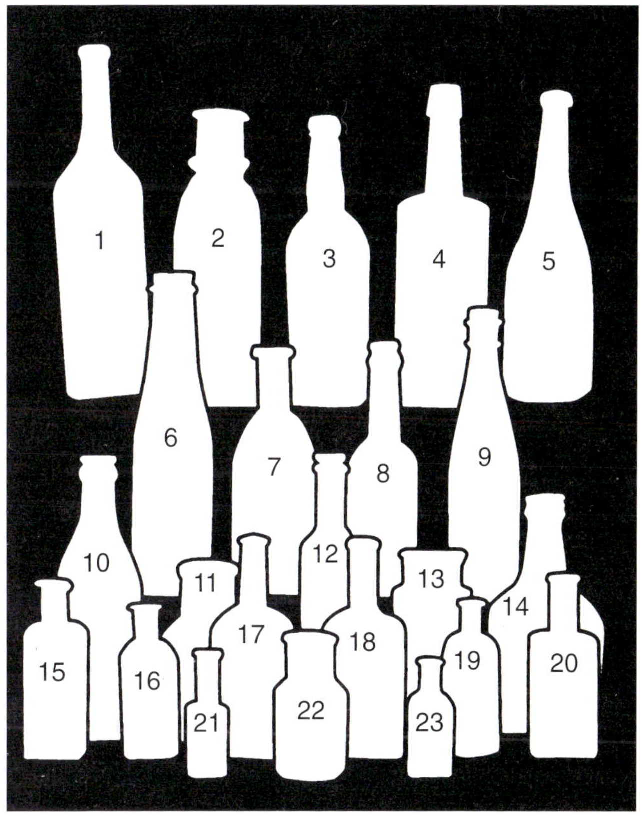

Plate 5. Glass Bottles.

1. *Wine bottle, 1825-1875.*
2. *"Cathedral" style pickling jar, 1840-1890.*
3. *Ale bottle, 1860-1900.*
4. *J. R. Nichols oil/elixer bottle, 1860-1872.*
5. *Wine bottle, 1825-1875.*
6. *Ketchup bottle, 1929-1954.*
7. *Burnett's Cocoaine Hair Tonic bottle, 1847-1840.*
8. *Lea & Perrins Worcestershire, 1880-1900.*
9. *Ketchup bottle, 1929-1954.*
10. *Soda "Coal Belt" bottle, 1903+.*
11. *Unspecified food bottle, 1920-1945.*
12. *Square condiment, 1929-1945.*
13. *Pill bottle, 1860-1870.*
14. *Old Quaker whiskey bottle, 1929-1954.*
15. *Unspecified medicine bottle, 1870-1915.*
16. *Unspecified medicine bottle, 1870-1915.*
17. *Bluing bottle, 1865.*
18. *Bluing bottle, 1818-1860.*
19. *Unspecified medicine bottle, 1911-1929.*
20. *Unspecified medicine bottle, 1870-1915.*
21. *F. Hoyt & Co. perfume bottle, patent 1891.*
22. *Colgate & Co. powder bottle, 1870-1915.*
23. *Unknown function, 1860-1870.*

Plate 6. *Ceramic vessels and tableware.*

west corner of the structure. Additional artifacts recovered from the kitchen suggest many uses of the structure. Intact glass bottles recovered from the kitchen include fruit jar stoppers, ale and sarsaparilla bottles; bluing bottles for laundry activities; and pill, unknown medicine, and hair tonic bottles for medicinal and personal grooming storage (*Plate 5*). All of these activities could be attributable to either the Allen or Lodge occupations, while the final activity is most likely representative of the Dr. Lodge residency. It is known that Dr. Lodge ran his office in a structure north of the house. The detached kitchen would have functioned nicely as a storage facility for his medicines.

Artifacts recovered from the cold storage facility were also suggestive of storage activities as they consisted of cathedral pickle jars, wine bottles, ale bottles (*Plate 5*), and an ironstone pitcher and tureen (*Plate 6*). It is likely that beverages and pickled food would have been stored in the facility in order to keep them cool.

The summer kitchen likely dates to both the Allen and Lodge occupations. This assumption is based on diagnostic ceramic and glass artifacts recovered from within the foundation; diagnostic building materials; and archival research. The Willis Allen occupation of the home was so short in duration that it was highly unlikely that a feature representative solely of this period would be found. Mortar used in constructing the summer kitchen is similar to that found on the fireplace and much of the existing house. The mortar, however, used in constructing the west wall surrounding the fireplace and the north and south walls extending three feet east of the fireplace is different. This suggests that two construction episodes: one including the summer kitchen, fireplace and chimney, and much of the existing house; and the second one involving the walls mentioned above. The fireplace was originally attached to the summer kitchen and not the existing house. At some later time, perhaps during the Augustus Lodge occupation, renovative work connected the fireplace with the existing house. This may have been undertaken in response to a structurally deteriorating summer kitchen since, by the late 1930s (the source of the painting of the house), clearly the summer kitchen is gone and the fireplace and chimney are attached to the house.

Research on eighteenth century detached kitchens of the Tidewater region in Virginia has identified regularized dimensions and uniform construction features. Common features include wooden construction, one story gable roof with a loft, one or two doors, windows, a large hearth, dressers, and a bake oven. Additional interior features such as cellars and ancillary spaces were not discussed. Generally these kitchens came in one of two varieties: one room with one or two gable-end chimneys; or two rooms with a central chimney and two hearth openings. Detached kitchens were generally either square or rectangular in shape and comprised anywhere from 400 to 800 square feet in area.

Detached kitchens were rarely within 20 feet of the living quarters, depending of course on available space. Detached kitchens were located away from the house for a number of reasons including odors, smoke, heat, and fire hazard. Activities commonly performed in detached kitchens other than cooking includes food preparation, laundry, storage, and sleeping. Detached kitchens often served as the living quarters for slaves or hired help in Virginia and may very well have been the case at the Allen House as well.

The detached kitchen at the Allen House diverges somewhat from the features commonly associated with eighteenth century Virginia kitchens. Firstly, the summer kitchen appears to have been constructed of brick and not wood. This is surmised from several sources. While only the brick foundation was recovered, the near absence of nails recovered from the summer kitchen would suggest that it was not built of wood. A wooden frame structure would have left many more nails than what was noted in the summer kitchen. Two areas of brick fall outside of the foundation, one to the west and one to the north of the structure, would indicate that brick was used in the wall construction as well. Prior to this investigation, a brick patio was located in the backyard that was undoubtedly constructed of bricks from the summer kitchen. It was noted in the discussion of eighteenth century detached kitchens that brick replaced wood as a common building material in the nineteenth century.

A second difference between eighteenth century Tidewater Virginia summer kitchens and the detached kitchen documented at the Allen House includes the proximity of the Allen House kitchen to the existing house. Tidewater kitchens were rarely located within 20 feet of the house, yet the Allen House kitchen is located, at most, seven feet from the house, and if the fireplace and kitchen were connected, much closer than that. Certainly space could not have been the reason as during the Allen and Lodge occupations, the lot was located outside of the city limits of Marion. It appears that perhaps access to the kitchen was more important than the unpleasant attributes cited in the Tidewater example, such as odors, smoke, heat, and fire hazards.

A final difference involves the size of the kitchen. Tidewater kitchens on average ranged in size from 400 to 800 square feet, while the Allen House kitchen was 144 square feet. This substantially smaller size would suggest that perhaps the Allen House kitchen did not serve as sleeping quarters but merely a work station. It may also indicate a different socioeconomic status of the Tidewater occupants and the Allens and Lodges. Maybe a larger kitchen was required by the Tidewater occupants to facilitate feeding a larger work force and preparing for large scale social events. The differences may also reflect changes through time and space. Perhaps further research will show that detached kitchens became smaller and less important through time; or that detached kitchens were less substantial in the Midwest than in the Southeast. Such research questions can and will be addressed through further historical and archaeological research.

It is likely that the summer kitchen was used by both the Allens and the Lodges. The construction of the kitchen is consistent with the construction of the house; artifacts found in the kitchen and cold cellar date to the mid to late 19th century which covers the time span of the Allen and Lodge occupations; and archival evidence suggests that the kitchen was gone by the time the Boles moved in.

Plate 7. East view of excavated cistern at northwest corner of house.

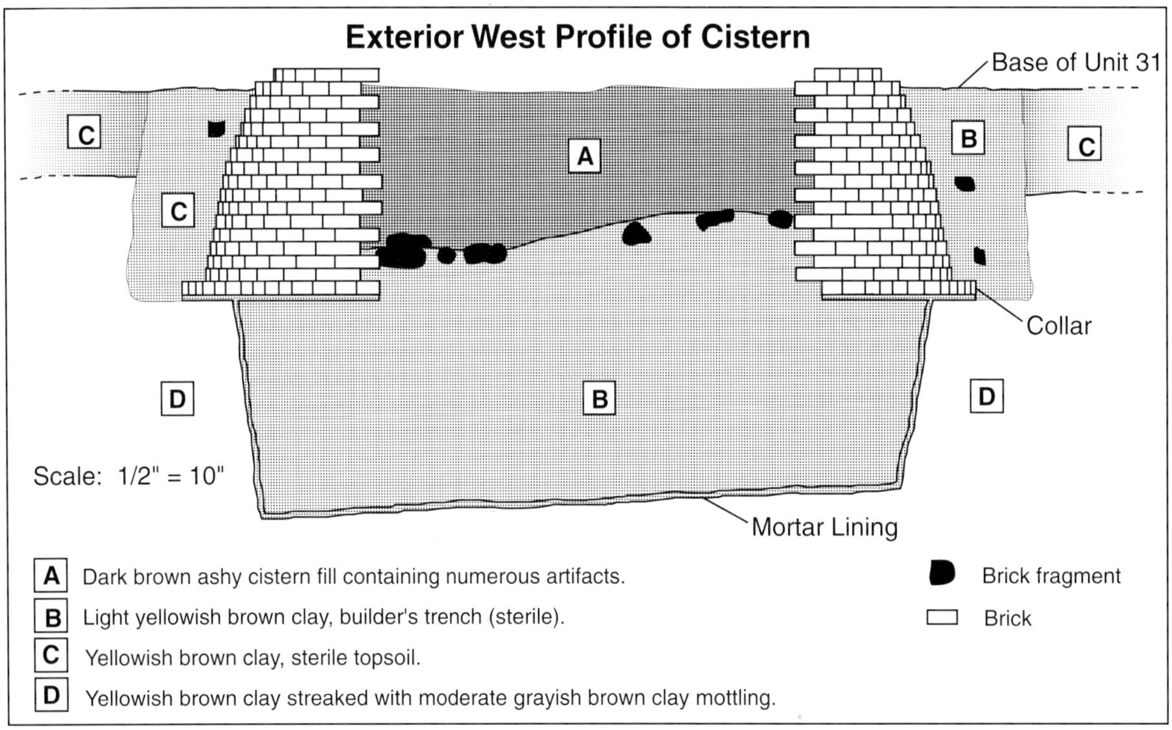

Figure 3. East profile of cistern at northwest corner of house.

Cistern

The first evidence of the cistern was identified as a brick arc in the corner of a 2 ft. by 2 ft. test unit excavation approximately six inches below the ground surface. The unit was eventually expanded to uncover the entire cistern perimeter (*Map 2*). The cistern was circular in shape measuring seven and a half feet in diameter and was located about six and a half feet northwest of the northwest corner of the recent addition to the 1854 house (*Plate 7 and Figure 3*). The cistern was constructed of handmade soft mud brick and sand mortar. The cistern was collared approximately three feet below the orifice and measured 10 ft. 8 in. in diameter. A builders trench extended 11.5 ft. in diameter and 5.5 ft. below the cistern and was filled with yellowish brown clay. Dark brown ashy soil that was filled with artifacts extended about two feet below the orifice. At that point a layer of large brick fragments was encountered. Below the brick fragments the yellowish brown clay, identical to the builder's trench fill, extended down 3 ft. 6 in. to a sandy mortar-lined floor. The floor measured 8 ft. 4 in. in diameter at which point sandy mortar walls expanded upward to meet the interior of the cistern collar.

The sandy mortar lining was used to waterproof the bricks. The profile of the cistern, however, shows the mortar lining extending well below the brick. This is a puzzle. Why was the brick not used to line the cistern below the collar? Why was so much of the cistern filled in with sterile yellowish brown clay? In an attempt to answer these questions we must refer to nineteenth century cistern construction techniques.

Cisterns built in the nineteenth and early twentieth centuries were generally constructed of one of the following materials: stone, brick, wood, or poured concrete. Cisterns also could be unlined or simply plastered. References to plastered and unlined cisterns indicate that this form of cistern was built primarily before 1850. Brick-lined cisterns, on the other hand, date to the second half of the nineteenth century (1850-1900). Poured concrete was the preferred building material during the twentieth century. The profile of the Allen House cistern suggests that two construction methods were used, brick and plaster. This conflicts somewhat with the construction dates given above and it does not address the problem of the sterile yellowish brown clay fill. A second interpretation would have an existing plastered cistern from a previous occupation being renovated with the construction of the Allen House in 1854. In other words, an earlier mortar lined cistern predated the brick-lined cistern which was built on top of it. The scenario would be as follows: a cistern whose walls were plastered with a sandy mortar was excavated to a depth of 5 ft. 6 in. and a diameter of 8 ft. 4 in. at its base; at some later point in time, the cistern was filled in to a depth of 3 ft. and a brick collar and lining was constructed; the brick walls and unlined floor were plastered with a similar sandy mortar; and finally the filling in of the cistern once it was no longer functional. This final stage would be when the site inhabitants would use the empty cistern for garbage (artifacts) disposal. This garbage is then represented by the approximately two feet of dark brown ashy fill located in the cistern.

Eric Schroeder documented an extensive list of references concerning the construction of cisterns during the nineteenth century. The following is an extended quote from The Prairie Farmer (1843) which details the plaster method of construction:

To make a cistern of about 40 barrels–that being suitable size for a common family–I would commence by digging a circular hole, 8 feet in diameter and two feet deep; having made the bottom tolerable level, strike a circle 6 1/2 feet in diameter on it, this leaving a shoulder all round to rest the plank on. Then dig by the circle a hole 6 feet deep, drawing gradually in as you go down, so that at the bottom it will be about 5 1/2 feet diameter; then round out the bottom, about 18 inches deeper, thus making it somewhat in the shape of a pot ash kettle, but considerably deeper in proportion.

The walls and floor of the hole were then plastered with as many as three coats of cement. Oak planks would then be laid on top of the hole; the cracks between the planks would be plastered; and then the whole thing would be covered with earth to protect the water from frost. While the dimensions cited in the extended quote differ somewhat from that noted at the Allen House, the general process corresponds nicely with what is identified in the profile. A plaster-lined cistern existed before the brick-lined cistern. The shoulders where the planks would have rested are easily identifiable in profile at about 3 feet below the original surface. The cistern extended down to about 5 1/2 feet below the original ground surface. It is likely that the plastered cistern was still functional at the time the brick cistern was constructed because there are no artifacts (garbage disposal) associated with the mortar-lined portion below the brick collar. Consequently the original builder's trench and much of the cistern interior were filled with clay at the time of the brick cistern construction. Perhaps the extent of filling in the cistern corresponds with the location of the brick collar at about 3 feet below the original surface. By the 1850s unlined and plaster-lined cisterns were losing favor because of their susceptibility to rodent damage and consequently contamination. Perhaps for that reason only the mortar-lined portion of the cistern was filled and a brick cistern was constructed with the construction of the house in 1854.

The cistern most likely collected rain water off the roof of the wooden addition of the home. This soft water was used primarily for washing, although consumption may also have been important. Cisterns were generally filled in with the garbage once their usefulness had diminished. Consequently they are excellent places to find a wide variety of artifacts. Such was the case at the Allen House.

A total of 211 artifacts were recovered from the cistern. These include 8 intact or nearly intact ceramic vessels and 11 intact glass vessels. Additional artifacts include a bone toothbrush, ceramic pipe bowls, clay marble, a porcelain doll tea plate, porcelain and metal furnishings, and butchered pig and cow bone. The complete ceramic vessels include 1 whiteware plate, 1 whiteware pitcher, 1 ironstone pitcher, 1 whiteware cup, 1 whiteware shaving mug, 1 ironstone serving bowl, 1 stoneware jug, and 1 stoneware churn (*Plate 6*). All of the ironstone and most of the whiteware was manufactured in England. Only the shaving mug and stoneware jug and churn were manufactured in the United States. The intact glass vessels consisted of 6 bluing bottles, 3 medicine bottles, and 2 powder bottles. A mean ceramic date of 1877 and a mean glass date of 1889.8 securely date the disposal of garbage in the cistern to the Augustus Lodge occupation. Therefore the cistern was most likely used during the Allen and perhaps the first half of the Lodge occupations.

Well

The well was discovered once the existing frame garage was removed (Map 2). The underlying well measured roughly 4 1/2 ft. in diameter and extended to a depth of 9 1/2 ft. below the well orifice. At that point water was encountered. The well was constructed of machine made brick and sandy mortar. It is known that the well predates the construction of the garage in 1945. It is also likely that the well predates a 1937 Sanborn Fire Insurance Map which fails to identify a well at that location. Machine-made bricks were not commonly used until after 1890 and the first machine-made brick factories in Marion date to 1898 and 1903 respectively. Therefore the well could not have been constructed prior to 1890 and, if the brick was purchased locally, was more than likely constructed around the turn of the century.

A total of 32 glass bottles were recovered from the well. These bottles functioned in kitchen and household related activities. The kitchen related bottles include ketchup, sauce, foodstuff, whiskey, and coffee, while the household related bottles include bleach, medicinal, facial cream, and lotion. An average date based on known manufacture dates of these artifacts is 1930.

Wells, as with cisterns, were used for garbage disposal once their primary function had changed. The cause for this change could be due to a variety of reasons including contamination of the water; lack of water; or perhaps the introduction of a preferred water source, such as a city waterline.

Approximately 25% of the well was excavated.

A profile of the excavation reveals, in layer cake fashion, the history of the disposing of garbage in the well. Six zones, or layers, were identified. Each was distinguished on the basis of soil color, soil texture, and/or distinctive artifacts. The bottommost layer consisted of a dark ashy silt that was filled with glass bottles and some ceramic artifacts. As this is the lowest zone, it represents the first episode of filling in the well. In turn each successive zone represents a more recent deposition with the top zone being most recent. As mentioned above, it is known that the top zone dates to 1945, the year the frame garage was built over the well. Bottles recovered from the bottom zone date, on average, to 1930. One bottle recovered from this zone, however, has as its earliest date 1933. Therefore the initial filling episode must have covered a period of time which, in part, postdated 1933.

The ashy nature of the soil suggests that perhaps a fire of some sort occurred at that time. The zone is relatively thick (approximately 4 ft.) and represents nearly half of the entire profile. An ash zone that thick with as many kitchen and household related bottles associated with it more than likely represents an extended period of deposition, rather than a relatively quick period associated with a burning episode. Perhaps ash from the household fireplaces, as well as possibly from the outdoor fireplace once associated with the summer kitchen, was routinely dumped in the old well along with other garbage.

The profile, however, documents five additional zones each of which are roughly a foot in depth. Interestingly two of the five zones are very ashy. Each of the ashy zones are separated by layers essentially devoid of ash. All of the layers contain cinders suggesting that coal burning activities were continuous. This may also be due to soil leaching or natural disturbance, the churning and stirring of soil by organisms such as worms or termites. In any case the profile suggests that during the initial deposition of ash in the well, ash was being produced year round. Subsequently ash was produced and deposited seasonally. At the point in time when the well was no longer used for water, the fireplace once associated with the summer kitchen must have still been used for cooking and/or burning trash. That would explain a continual production of ash which would in turn produce a thick layer of deposition in the well. In other words, the fireplace would have been periodically cleaned throughout the year and the ash would have been dumped. Once the outdoor fireplace was no longer used, ash would be produced and deposited seasonally. The occasional dumping of coal ash from a coal burning stove would account for the minor presence of cinders throughout the well fill. If it is true that the alternating layers of ash indicate seasonality, then it should be possible to reconstruct the number of years the well functioned as a garbage disposal area. Therefore alternating layers of ash/no ash suggest the number of years the well was used for disposal after the outdoor fireplace was no longer used and before the frame garage was constructed. It is therefore plausible that the outdoor fireplace was used as late as the early 1940s. The foregoing evidence suggests that the filling of the well occurred while the Tyner family lived in the house (1929-1945).

Plate 8. Displaced piers from outbuilding.

Outbuilding

The 1937 Sanborn Fire Insurance Map (*Map 3*) documents an outbuilding due west of the southwest corner of the house. The backhoe stripping of the sod in this general vicinity revealed several large sandstone blocks which most likely served as foundation piers for that structure (*Map 2 and Plate 8*). A total of 18 test units were excavated in order to define the sandstone pattern. An area of approximately 10 feet by 10 feet was identified. Four additional sandstone blocks were documented about five feet north of the area and probably represent displaced portions of the outbuilding.

Despite the intensive excavation, very few artifacts were recovered. A "Lea & Perrins" bottle and bottle stopper with averaged manufacture dates of 1890 and 1903 respectively, were recovered from test unit excavation. This corresponds nicely with the other diagnostic artifacts recovered from the area, wire nails. This type of nail replaced machine-cut nails in popularity by 1890. The vast majority of nails recovered from this feature were wire nails. This suggests that the outbuilding was constructed after 1890 and likely was wooden frame. This latter contention is supported by the absence of brick fragments, the abundance of nails, and the painting of the house which clearly depicts a wooden frame outbuilding at this location. Due to the relative absence of artifacts, little more can be said about how the outbuilding was used.

Trash Pit

A circular feature measuring 1 ft. 10 in. in diameter was discovered approximately four feet west of the southwest corner of the summer kitchen and about eight feet east of the southeast corner of the outbuilding (*Map 2*). A circular stain filled with metal fragments was noted after the sod had been removed by the backhoe. The feature extended 6 inches below the scraped surface and contained tin can parts and glass bottles. The glass bottles included 3 sauce bottles, a soda bottle, and a juice bottle. The bottle manufacturing techniques were diagnostic of the early twentieth century. It is likely that this feature functioned as a shallow trash pit that was probably associated with the outbuilding.

Old Porch/Family Room and Pantry

As mentioned earlier, the old porch was converted into an enclosed family room by James and Lorene Wilson, who resided in the house from 1964 to 1973. In addition the old brick-walled pantry was converted to a powder room. The painting of the house from the late 1930s (*Figure 1*) shows how the porch appeared at an earlier date. Both Sanborn Fire Insurance Maps also revealed the simply constructed frame addition (*Figure 2*). Removal of this section for the new addition revealed several displaced sandstone blocks which may have served as footers for the original porch. One sandstone block, located in the southwestern corner, may have actually served as a pier for a post which appears in the old painting (*Map 2*). Unfortunately, no artifacts were recovered from this area.

Postmolds

Three postmolds were discovered just south of the garage and west of the cistern (*Map 2*). The postmolds were oriented in a north-south line and varied in size from 8 to 10 inches squared. No other postmolds were discovered near them; however, other molds may have been destroyed by previous impact. The postmolds may have been part of an outbuilding. Since only the one outbuilding appears on the painting (*Figure 1*), the structure that the postmolds are part of may be older than the one we know about.

Conclusions

As archaeologists it is gratifying to encounter people like Becky and Bernie Paul who are not required by state or federal regulations to preserve the archaeological resources in their own backyard, but who care enough to commission archaeological investigations before valuable historical and archaeological information is destroyed by construction activities. The Allen house investigations form building blocks for wider historical reconstructions in southern Illinois and will be compared with investigations from rural and urban archaeological sites as a means of studying changing lifeways. The Willis Allen family lived on the near outskirts of Marion and were buried in their own backyard, an uncommon practice today. Subsequent owners of the house made the change from well water to indoor water, from outhouses to indoor toilets, from candles and kerosene to electricity and the many other conveniences of modern living. Like the natural environment, it is important that we preserve our past cultural environments as well so that we can place the lives of our ancestors in these everyday settings.

BIBLIOGRAPHY

Erwin, Milo, The History of Williamson County, Illinois. Williamson County Historical Society, Marion, Illinois, 1876.

Gurcke, Karl, Bricks and Brickmaking: A Handbook for Historical Archaeology. The University of Idaho Press, Moscow, Idaho, 1987.

Hubbs, Barbara Burr, Pioneer Folks and Places: An Historic Gazetteer of Williamson County, Illinois. Williamson County Historical Society, Marion, Illinois. Reprint. Originally published 1939 by the Herrin Daily Journal, Herrin, Illinois, 1979.

A Lady's Friend, "Cisterns". The Prairie Farmer, 6 (4), 1846.

Leach, Detached Kitchens of the Tidewater Region. Collegiate Press, Texas A & M University, 1986.

Mason, Roger D., "Euro-American Pioneer Settlement Systems in the Central Salt River Valley of Northeast Missouri". Publications in Archaeology No. 2. University of Missouri, Columbia, 1984.

McCorvie, Mary R., Mark J. Wagner, Jane K. Johnston, Terrance J. Martin, and Kathryn E. Parker, Phase III Archaeological Investigations at the Fair View Farm Site (11-Sa-336): A Historic Farmstead in the Shawnee Hills of Southern Illinois. Cultural Resources Management Report No. 135 ed. American Resources Group, Ltd., Carbondale, IL, 1989.

Mitchell, R. D., Commercialism and Frontier: Perspectives of the Early Shenandoah Valley. University Press of Virginia, Charlottesville, 1977.

Newton, M. B., "Cultural Preadaptation and the Upland South". Man and Cultural Heritage: Papers in Honor of Fred B. Kniffen, edited by H. J. Walker and W. G. Haag, pp. 143-154. Louisiana State University, Baton Rouge, 1974.

Prairie Farmer, The, "Review of a Prairie Farmer for April, 1843, Cemeter Cisterns". The Prairie Farmer, 3 (5):92-93, 1843.

Schroeder, Erich K. (editor), "Landscape, Architecture, and Artifacts: Historical Archaeology of Nineteenth-Century Illinois". Illinois Cultural Resources Study No. 15, Illinois Historic Preservation Agency, Springfield, 1991.

Smith, B., "Cement Cisterns". The Prairie Farmer, 3 (6), 1843.

Wilcox, J. F., Historical Souvenir of Williamson County, Illinois. The LeCrone Press, Effingham, 1905.